We should not eat a lot of
fats, **oils** and **sweets**.

Oil is a kind of fat.

Sweets have a lot of sugar
in them.

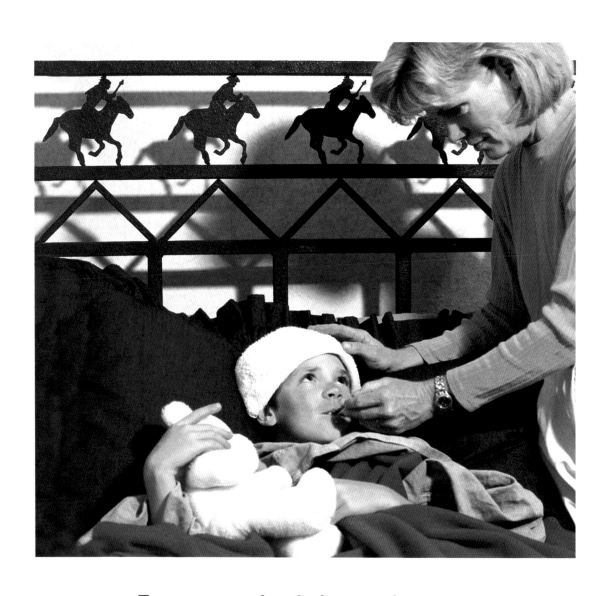

Too much fat and sugar
can make us ill.

Too much fat and sugar
can be **unhealthy**.

There are many snacks in
this group.

We should not eat too
many biscuits.

We should not eat too
many chips.

We should not drink too
many fizzy drinks.

We can choose healthy snacks.

We can eat popcorn.

We can eat vegetables.

We can eat fruit.

We can eat yogurt.

Watching what I eat keeps
me healthy.

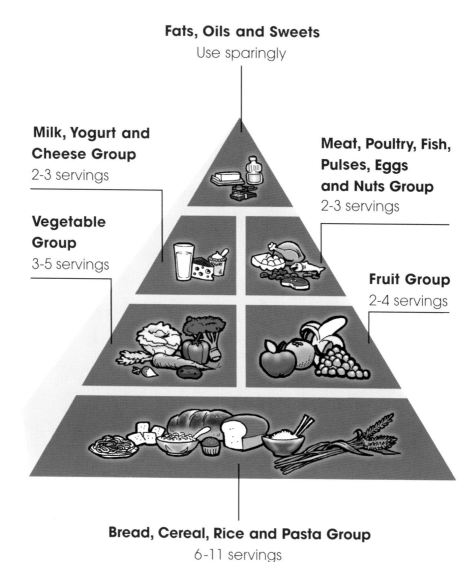

Fats, Oils and Sweets
Use sparingly

Milk, Yogurt and Cheese Group
2-3 servings

Meat, Poultry, Fish, Pulses, Eggs and Nuts Group
2-3 servings

Vegetable Group
3-5 servings

Fruit Group
2-4 servings

Bread, Cereal, Rice and Pasta Group
6-11 servings

18

Fats, Oils and Sweets

The food pyramid shows us how many servings of different foods we should eat every day. Fats, oils and sweets are at the top of the food pyramid. This part of the pyramid is the smallest because you should not eat very many foods from this group. Many snacks like biscuits, chips, chocolate and fizzy drinks belong to this group. The foods in this group taste good, but they have a lot of sugar or fat in them.

Fats, Oils and Sweets Facts

 The most popular ice cream flavour is vanilla.

 The first chocolate chip cookie was invented by Ruth Graves Wakefield in 1930.

 Crisps are the most popular after-school snack in Britain.

 The custard cream is the most popular biscuit in Britain.

The first Easter egg was made in 1873.

 The British are the biggest chocolate eaters in Europe! We eat an average of 10 kilograms a year.

Glossary

 fats – parts of food that give you energy

 healthy – not sick; well

 oils – fatty liquids used in food

 sweets – foods that contain a lot of sugar and taste good

 unhealthy – sick; not well

Index

biscuits – 9, 19

chips – 10

fizzy drinks – 11, 19

fruits – 15

popcorn – 13

sugar – 5, 6, 7, 19

vegetables – 14

yogurt – 16

This book was first published in the United States of America in 2003.

First published in the United Kingdom in 2008 by
Lerner Books,
Dalton House,
60 Windsor Avenue,
London SW19 2RR

Website address: www.lernerbooks.co.uk

This edition was updated and edited for UK publication by Discovery Books Ltd., Unit 3, 37 Watling Street, Leintwardine, Shropshire SY7 0LW

Words in **bold** type are explained in the glossary on page 22.

British Library Cataloguing in Publication Data

Nelson, Robin, 1971-

    Fats, oils and sweets. - (First step non-fiction. Food groups)
    1. Lipids in human nutrition - Juvenile literature
    2. Sugars in human nutrition - Juvenile literature 3. Oils
    and fats, Edible - Juvenile literature 4. Sugars - Juvenile
    literature
    I. Title
    641.3

    ISBN-13: 978 1 58013 388 3

The photographs in this book are reproduced through the courtesy of: © Todd Strand/Independent Photo Serice, front cover, pp 3, 4, 5, 8, 9, 11, 12, 13, 15, 17, 22 (top, middle, second from bottom); © PhotoDisc/Getty Images, pp 2, 7, 10, 22 (second from top); © Royalty-Free/CORBIS, p 6, 22 (bottom); © Jorge Delgado/Istockphoto.com, p 14; © Midwest Dairy Association, p 16.

The illustration on page 18 is by Bill Hauser/Independent Picture Service.

Printed in China